Pioneering Ocean Depths

OCEAN
DEPTHS

S A N D R A   M A R K L E

Atheneum Books for Young Readers

ACKNOWLEDGMENTS

The author would like to thank the following people for sharing their expertise and enthusiasm: Dr. Robert R. Hessler (Scripps Institution of Oceanography), Marcia Allen (Natural Energy Laboratory of Hawaii), Kathleen K. Newell (University of Washington, School of Oceanography), Dr. James Ledwell (Woods Hole Oceanographic Institution), Dr. Craig Smith (University of Hawaii, Department of Oceanography), Dr. Wilford D. Gardner (Texas A&M University, College of Geosciences and Maritime Studies), and Dr. Gerald H. Krockover (Purdue University).

PHOTO CREDITS

Cover: Norbert Wu; 2–3: Russ McDuff; **p. 5** E. A. Widder; **p. 6** NOAA National Geophysical Data Center; **p. 8** American Petroleum Institute; **p. 9** Ken C. MacDonald; **p. 10** Rob McDonald; William A. Markle, Compu-Quill; **p. 11, 12** University of Washington, School of Oceanography; **p. 13** Ocean Drilling Program/Texas A&M University; **p. 14** University of Washington, School of Oceanography; **p. 15, 16** Rob McDonald; **p. 17** U. S. Coast Guard; **p. 18** Norbert Wu; **p. 19** Harbor Branch Oceanographic Institute/Tom Smoyer; **p. 20** Rob McDonald; **p. 21** Deep Submergence Unit, U. S. Navy; Woods Hole Oceanographic Institution; **p. 22** Sandra Markle; **p. 23** Digital images by Peter W. Sloss, NOAA, National Geophysical Data Center; **p. 24** Robert Tyce, University of Rhode Island, Department of Oceanography; **p. 25** Shef Corey, Ocean Mapping Development Center, University of Rhode Island; **p. 26** Wilford Gardner, Texas A&M University; **p. 27** University of Washington, School of Oceanography; Wilford Gardner, Texas A&M University; **p. 28** William R. Bryant, Texas A&M University; Rob McDonald; **p. 29** Ocean Drilling Program/Texas A&M University; **p. 30** Norbert Wu; **p. 31** Norbert Wu; P. J. Herring, Institute of Oceanographic Sciences, Surrey, United Kingdom; **p. 32** Harbor Branch Oceanographic Institute/Tom Smoyer; Rob McDonald; **p. 33** Dudley Foster, Woods Hole Oceanographic Institution; **p. 34** V. Tunnicliffe, University of Victoria, Canada; **p. 35** John Porteous; V. Tunnicliffe, University of Victoria, Canada; **p. 36** Craig R. Smith; Helmut Kukert; **p. 37** W. Kenneth Stewart, Woods Hole Oceanographic Institution; **p. 38** American Petroleum Institute; **p. 39** EXXON; **p. 40** Margaret Sulanowska, Woods Hole Oceanographic Institution; **p. 41** Wilford Gardner, Texas A&M University; **p. 42** Silvia Mercoli and Kenneth Rinehart; **p. 43** NASA; James R. Ledwell, Woods Hole Oceanographic Institution; **p. 44** Rob McDonald; **p. 45, 46** Greg Vaughn; **p. 47** Harbor Branch Oceanographic Institute/Tom Smoyer.

The text of this book is set in 11-point Trump Mediaeval.
Printed in the United States of America.
10 9 8 7 6 5 4 3 2 1

LIBRARY OF CONGRESS CATALOGING-IN-PUBLICATION DATA

Markle, Sandra.
    Pioneering ocean depths / Sandra Markle.—1st ed.
      p.   cm.
    Summary: The ways scientists are learning more about the ocean and seafloor with sonar, submersibles, and other new techniques, and what it means for mankind.
    Includes index.
    ISBN 0-689-31823-5
    1. Underwater exploration—Juvenile literature.   [1. Underwater exploration.   2. Oceanography.]
I. Title.
GC65.M34   1995
551.46—dc20
                                                                       93-33555

For HOLLY with love.
Have a great sophomore year!

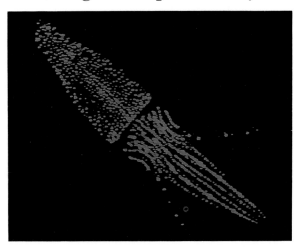

Into the Ocean Depths

From space, it's easy to see that the greatest part of the earth's surface is water. People have been studying these vast oceans since ancient times. Even with the help of modern technology, though, much of this watery realm remains mysterious and unexplored. This is especially true of the deepest parts—the earth's largest and most challenging frontier.

So how are scientists exploring this deep ocean frontier? What's down there? And how are the discoveries that are being made proving useful?

Just going into the ocean depths is a challenge. It's dark. Nearly 60 percent of the visible light entering the ocean is absorbed within the first 3 feet (0.9 meter); about 80 percent is gone after 30 feet (9 meters). A tiny bit of light penetrates down to about 500 feet (150 meters) if the water is very clear, but deep down in the ocean, it's very, very dark. Without the warmth created by the sun's energy, the deep ocean is also extremely cold. Furthermore, the weight of the seawater plus the air above it pressing down make the water pressure greater the deeper you go.

Come share the adventure and along the way perform investigations that will let you see what it's like to be a scientist exploring the ocean depths.

Opposite: This view of the land under the ocean was created by a computer, based on satellite data about wave height. Water actually piles up over seamounts and is lower where there are deep canyons. The earth is said to have four main oceans—Arctic, Pacific, Atlantic, and Indian. From this South Pole view, though, it's clear that all the water mixes, forming one big ocean system.

How Are Scientists Exploring the Depths?

They're Mapping with Sound Waves

To find the ocean's depth, early sailors took soundings by lowering a weighted line marked in 6-foot (1.8-meter) units called fathoms. By the early 1900s, scientists had developed sonar, which uses sound waves to measure the ocean's depth at any one point. Special instruments send out pulses of high-frequency sound waves that travel to the seafloor and are reflected back to the ship, where the echo is detected. By recording the depth of a series of points and plotting these on a graph, scientists can create a profile of the seafloor. Studies have proved that sound normally travels through seawater at the rate of 5,000 feet (1,500 meters) per second. So to determine the distance from the ship

In this special kind of sonar system, sounds emitted by air guns travel through the layers of sediment and rock on the ocean bottom before bouncing back. A special computer uses the length of time it takes the echoes to reach the surface to generate an image of the seafloor's structure. Geologists study these to detect formations where oil and gas are likely to be found.

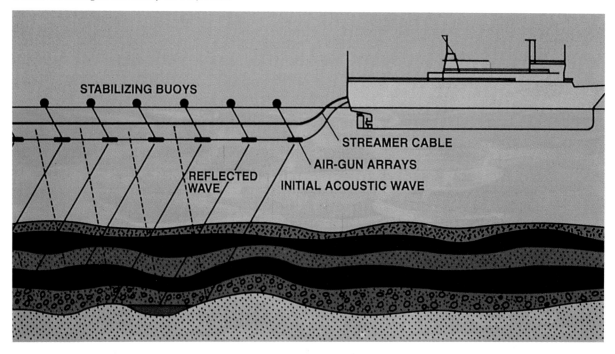

STABILIZING BUOYS

STREAMER CABLE

AIR-GUN ARRAYS

REFLECTED WAVE

INITIAL ACOUSTIC WAVE

The profile of these deep-sea volcanic mountains was produced using the Sea Beam sonar scanning system. The different colors represent different heights so that scientists can see at a glance how tall the mountains are. Some are almost 7,000 feet (2,100 meters) tall, but still well below the surface. In all, more than a thousand seamounts and volcanic cones were discovered about 600 miles (966 kilometers) northwest of Easter Island in the South Pacific in an area about the size of New York State—the greatest concentration of volcanoes that's been discovered so far.

to the bottom of the ocean, we measure the length of time it takes the sound to go and return and multiply by 5,000 (1,500); then that product is divided by 2. For example, if it takes 1.5 seconds to detect the echo, the ocean is 3,750 feet (1,125 meters) deep at that point.

Today, seafloor mapping is done using the Sea Beam system, a special type of sonar that sends out multiple sonar beams simultaneously, which makes it possible to map a whole swath of the seafloor in one pass. Computers transform the individual sonar readings into a three-dimensional model of the mountains, trenches, and other features that shape the seafloor.

9

FIND THE MYSTERY SEA BOTTOM

To discover how learning the depth at a series of points can let you "see" the seafloor, you'll need a shoe box, scissors, a 10-inch (25-centimeter) piece of string, a dab of modeling clay, a ruler, a marking pen, and enough sand or clay to fill the box half full.

Pour the sand or soil into the box and shape it into deep canyons and mountains, making sure none is taller than the sides of the box. Cut a half-inch (1.25-centimeter) wide slot down the length of the lid, and put the lid on the box. Number from one through twenty—or more if you have room—along the slot, spacing the numbers evenly. Draw lines on a piece of paper to make a graph like the sample (right). Tie a knot at one end of the string and mold clay over it, forming a tiny ball. Mark the string at 1-centimeter intervals. Lower the string—clay ball first—through the slot at number one. When you feel the ball touch bottom, pinch the string where it touches the edge of the slot on the box

lid. Then pull the line out of the box and record the depth on your graph. Check the example to see how that's done. Lay the string next to the ruler if you need to compute a fraction between two marks. Repeat, measuring and recording the depth at each numbered location along the slot. Finally, connect the points you marked on the graph to see the profile of your model seafloor.

The weights on early sounding lines were coated with wax. Examine your clay ball. What information do you think this method of remotely sensing the seafloor provided that sonar can't? Why do you suppose a line could not easily be used to measure depths as great as those mapped with sonar? (Clue: Think about how long the line would have to be and how long it would take to let it out and reel it back up.)

Extra challenge: Instead of sand or clay, have a friend place an object, such as a shoe, inside the box. If the object's something small like a toy car, double-sided tape may be needed to anchor it directly under the slot. Now take soundings of the model seafloor again, recording the depth at each number on a new graph. Can you guess the identity of the mystery object from the profile? Why do you suppose oceanographers found it hard to interpret what the seafloor was like based on a one-dimensional profile like the one made by the old line-sounding method?

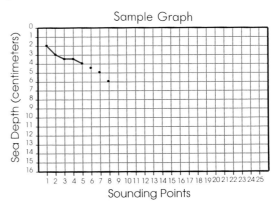

They're Collecting Seafloor Samples

Just knowing the shape of the seafloor isn't enough for oceanographers. They also want to study what's down there firsthand. The ocean depths, though, are as alien an environment for people as the surface of the moon. So scientists use special tools like the ones below and on the next page to collect samples and bring them up to the surface.

Have you ever stuck a straw straight down into a drink, put your finger over the top, and then lifted it up full of liquid? That's how this gravity corer works. The pipe has a sharp cutting end that's driven into the sediment by the pipe's own weight or by a piston that pounds it in. The pipe fills with seafloor material as it descends through it. Each sample, called a core, is about 30 feet (9 meters) long and 8 inches (20 centimeters) in circumference. When a core is first brought up on deck, it's encased in a plastic sheath. This sheath is marked to show what is the top, and it's labeled to indicate the area of the seafloor it came from. Then the core is divided into 5-foot (1.5-meter) lengths, and each section is split in half—half to be stored, half to be studied by scientists on board the research vessel.

Above: This special tool is lowered open and then triggered to snap shut when it strikes bottom. Here you can see what was collected when it took a bite out of the seafloor.

Right: Where it isn't quite so deep, a basketlike dredge is pulled across the bottom. Can you guess why the edge of the opening has sharp teeth?

WHAT'S DOWN THERE?

To see for yourself what it's like to collect samples over a long distance using special tools, sprinkle twenty pebbles or marbles, twelve paper clips, and ten balls of modeling clay about the size of the end of your little finger on the surface of your model seafloor. Then put the lid on the box and use tweezers or chopsticks to collect samples from two different numbered locations. Was it easy to collect your sample this way? Was what you collected really representative of what you know was on the model seafloor? If you collected any of the clay balls, you probably also discovered that the collection process can alter and even damage the sample.

Drilling a core sample in the deepest parts of the ocean is expensive and difficult. So nineteen countries joined together for the Ocean Drilling Program. This is the latest phase of a deep-sea drilling project that began in 1968.

Once a core sample is brought on board ship, paleontologists, scientists who study past life, examine the core to check for plant and animal fossils that will help them identify the age of the oldest material in the sample. Other scientists study the sediment's density, composition, radioactivity, and ability to conduct heat. Sensors are also lowered into the hole the core left behind to collect more information about the physical and chemical properties of that part of the seafloor.

This thin section of a rock collected from the seafloor reveals its crystal structure. Seeing this structure helps scientists learn something about what conditions were like on the earth when the rock was being formed.

This one-of-a-kind drill ship is called the JOIDES *(Joint Oceanographic Institutions for Deep Earth Sampling) Resolution. It can drill in waters up to 5 miles (8 kilometers) deep and extend almost 6 miles (9.6 kilometers) of drill pipe to obtain core samples. Twelve powerful thrusters guided by computers hold the ship steady and in position during the drilling process. Twelve laboratories equipped with state-of-the-art instruments aid scientists in analyzing the cores.*

They're Collecting Deepwater Samples

If you've ever been swimming and noticed the water felt different temperatures at different depths, it probably won't surprise you to learn that the amount and types of chemicals present in the ocean vary at different depths and at different locations.

Each of the special bottles in the picture is called a Nansen Bottle for the Scandinavian oceanographer who invented it in the 1800s. It's what oceanographers use to collect seawater from different depths so they can analyze its chemical makeup. When a measured line shows the Nansen Bottle has reached the desired depth, another line triggers a counterweight or spring-loaded system that snaps a lid shut, trapping the water that's inside. Besides collecting water samples, a probe in the center of the cluster of bottles has special instruments that measure temperature, pressure, and salinity, the amount of salt in the water.

This is a cluster of Nansen Bottles. Because these collectors still only provide a sort of snapshot look at the ocean at one depth and location, scientists sometimes send down instruments capable of reporting information steadily for months and even years.

To see for yourself how water can be collected from a specific spot underwater and brought to the surface for analysis, you'll need a quart jar, red and green food coloring, a plastic straw, and a white paper towel. Fill the jar full of cold tap water—the colder the better because the water molecules move more slowly and there's less natural mixing when the water's cold. Let the jar sit for a couple of minutes so the currents created by filling the jar slow down. Next, drop one drop of green food coloring into the water. Holding a finger over the top end of the straw, lower the other end past the sinking green coloring into the clear water at the bottom of the jar. Lift your finger for just an instant. Press down again and lift the straw out of the jar. Lift your finger to let the water inside the straw drain out on the paper towel. The wet spot will be clear, proving you were able to capture a little bit of water at the precise location you chose. Wait a minute for the green coloring to spread throughout the water. Then add a drop of red food coloring to the water. As soon as the red coloring sinks about halfway, repeat your collection process, but this time capture some of the bright red water. You should see the red-colored water move into the straw when you lift your finger off the end.

HOW SALTY IS SEAWATER?

Oceanographers check the salinity or saltiness of seawater at different locations and at different depths to try and understand how the water is circulating within the oceans. They hope to discover patterns of movement between the surface and the depths and from one ocean to another. The special instrument used to measure salinity is called a salinometer. It actually measures how well the seawater will conduct an electric current because the saltier the water is, the stronger the current it will conduct.

While you won't be able to measure differences in salinity, you can see for yourself that salty water conducts an electric current. You'll need aluminum foil, transparent tape, scissors, a 2.5-volt flashlight bulb, two D-cells (batteries), a measuring spoon, table salt, two pennies, and a glass pie plate.

First, make two foil wires about 12 inches (30 centimeters) long by placing tape on the dull side of the foil, cut out the taped strips, and fold each in half lengthwise. Stack the two D-cells so the knob end of the bottom one is touching the flat end of the top one. Wrap tape around the D-cells to hold them together.

Next, set up a system you know will light the bulb. To do this, set the flat end of the stacked D-cells on one foil wire. Wrap the other end of this wire around the bulb's screw base, and touch the tip of the bulb's base to the knob end of the battery. If the bulb doesn't light, check your connections. Make sure the two D-cells are pushed tightly together. Or you may need a new bulb or new D-cells.

Once the bulb lights, fill the pie plate two-thirds full of water. Set the flat end of the stacked D-cells on one of the wires, placing the free end of that wire in the water. Set a penny on it to hold the end underwater. Place one end of the other wire in the water—very close to but not touching the first wire. Anchor it with a penny. Wrap the free end of this wire around the bulb's screw base and touch the tip of the base to the knob end of the battery as before. The bulb probably won't light; if it does, the glow will be dim.

Remove the wires from the water, spoon in two tablespoons of salt, and stir until the salt is dissolved. Repeat the test. The bulb should glow this time. The grains of salt have broken down into sodium ions and chlorine ions. An ion is formed from an atom that has lost or gained electrons. Sodium ions have a positive charge while chlorine ions have a negative charge. When you put the wires connected to the two ends of the D-cell in the salty water, the ions move toward opposite charges—the sodium ions move toward the flat end of the battery because it has a negative charge and the chlorine ions move toward the knob end of the battery because it has a positive charge. These moving charged particles create an electric current that makes the bulb glow.

Wonder why saltwater is salty? Water seeping through sediment and running across rocky soil dissolves sodium chloride, or table salt, sulfate, magnesium, calcium, potassium, and traces of other substances that are collectively called salts. Rivers carry the dissolved salts and deposit them into the sea. Rain also carries particles from the air into the sea. Some of these chemicals are removed from seawater by animals, such as corals, as part of their normal life processes. Sea spray removes some too. Because just about as much salt is gained as is lost, the salinity of seawater remains about the same everywhere in the world. It's those subtle differences in saltiness detected by a salinometer that provide oceanographers with clues about the movement of seawater within the ocean.

Seawater is the least salty in polar regions because melting ice adds freshwater. It's saltiest in tropical regions around the equator because heat causes surface water to evaporate, leaving the salts behind.

They're Taking a Close Look

Because they know that the best way to study something is to actually see it for themselves, scientists who want to study the seafloor go there in submersibles designed especially for deep diving. It takes up to two hours to make the long descent, and two scientists, a pilot, and all their equipment must ride inside a sphere only 6 feet (1.8 meters) in diameter! At first, it's hot and stuffy inside the submersible. But as the vessel descends, the water outside gets colder and colder. This makes the inside of the submersible chilly too. The air pressure, though, is maintained at exactly the same level it is on the surface, making it possible to breathe easily.

While the deep-sea world outside is largely silent, the submersible is filled with noise. The carbon-dioxide scrubbers that recycle the air and make it breathable hum; so do the vessel's motors and electronic systems. The hull creaks as it adjusts to the increasing pressure outside. And the submersible's sonar navigation system makes pings that one scientist said sound like two-hundred-pound canaries peeping.

This submersible is making one of several dives to explore one specific area on the seafloor. Before it makes the first dive, special sound beacons, which pick up a signal sound and send one out at a different frequency, were dropped, creating a sort of net. By knowing where these beacons are located and computing how long it takes a signal sound to go and return, the pilot can precisely fix his position.

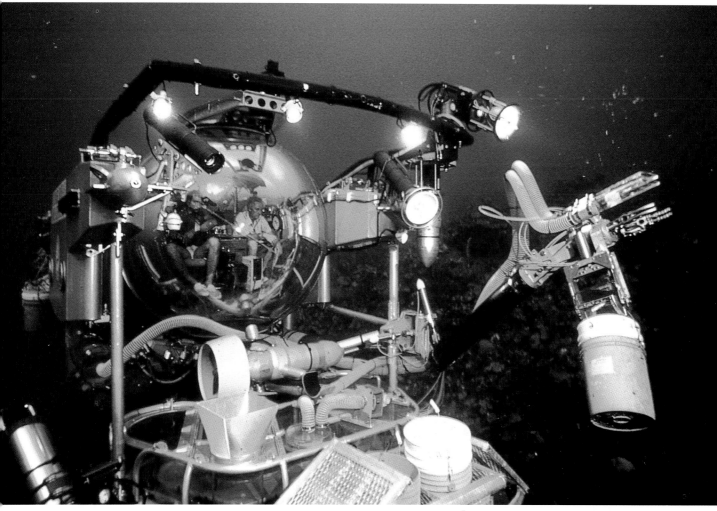

See what looks like a tube being held by the submersible's manipulator arm? Like cutting biscuit dough, this special corer is used to collect sediment that the scientists on board want to carry back to the surface for a hands-on examination.

Once on the seafloor, the scientists quickly forget any discomfort in the excitement of exploring this world few people ever get to see. Because no sunlight penetrates to the ocean's deepest realm, the outside of the submersible is studded with lights aimed in every direction. These are able to illuminate an area up to 40 feet (12 meters) away. To a geologist focusing on the rock structure, the deep seafloor reveals stark, bare lava or sediment-covered terrain. To a marine biologist looking for life, much of the deepest regions is sparsely populated, but there are communities of animals occupying microenvironments. Because their time in the depths is limited, the scientists take photos—both stills and videos—record notes, and perform tests using the submersible's manipulator arm and remote sensors to test the water, collect rock and sediment samples, and even collect animals to study back on the surface. All too soon, because of the limited battery power, it's time to start the long trip back up to the surface.

BUILD A MODEL SUBMERSIBLE

You'll need a clear two-liter plastic drink bottle, a bowl, and an eyedropper. Fill the bottle full of water. Adjust the amount of water in an eyedropper so it floats in a bowl filled with water with just a bit of the squeeze top showing above the surface. Place the eyedropper in the bottle and screw on the bottle cap.

The eyedropper is your model submersible. To make it descend, gently squeeze the sides of the bottle. This increases the pressure inside the bottle, which compresses the air inside the dropper. Since the air takes up less space, more water is able to move into the eyedropper, increasing its weight, and it descends. When you stop squeezing, the air inside the bottle has more room and the pressure decreases. Then the air inside the dropper expands, pushing some of the water out. This makes the model lighter, and it rises.

Dudley Foster, an engineer and pilot of the submersible *Alvin* for many years, explained:

"To make the submersible sink, we carry a thousand pounds [450 kilograms] of steel in the form of four plates. We also open ballast tanks, letting them fill with water. Once on the bottom, we drop half the steel plates and pump enough water out of the tanks to make the submersible neutrally buoyant, meaning it remains suspended above the seafloor at the exact depth we want.

"Driving Alvin *is sort of like flying a helicopter only much slower. There are propellers to make the submersible go forward. But there are also six thrusters, propellers surrounded by a metal shield, that shoot water in a specific direction. Using these, the submersible can also be made to move up and down, go backward, and turn. It's very hard to make* Alvin *go sideways. When we're ready to come up, we drop the remaining steel plates, pump water out of the ballast tanks, and float up."*

Can you make your model submersible dive all the way to the bottom of the bottle? Is it easier to make it go up and down or from side to side?

They're Exploring with Unmanned Submersibles

Because it's dangerous, costly, and time-consuming to actually travel to the seafloor, scientists have found other ways to explore the deepest parts of the ocean. One approach is to use a Remotely Operated Vehicle (ROV), meaning that instead of being operated by a crew, the submersible is connected to the mother ship by a cable and controlled by a pilot on the ship moving a joystick, like playing a video game. Images from video cameras mounted on the ROV travel over fiber-optic cable to monitors, letting a large group of scientists and researchers share views of what the ROV "sees." They can also make the manipulator arm collect samples just the way they would if they were on board the submersible.

While a ROV makes it safer to study the ocean depths and lets more people get a firsthand view, it still can't perform investigations lasting more than a few days. The operating costs for the mother ship and crew are just too great. To solve this problem, engineers at Woods Hole Oceanographic Institution developed the Autonomous (working alone) Benthic (deep-sea) Explorer—

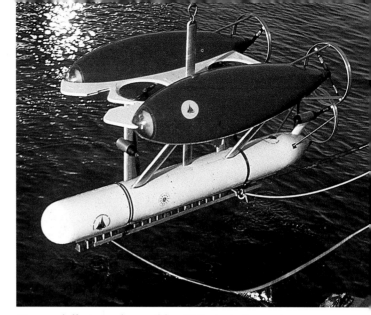

Now a full-size submersible, ABE started as a toy-size model in a two-story water tank. By working on something small, engineers could test different shapes for the submersible easily and inexpensively. In fact, ABE started out lima-bean-shaped, but that model twisted as it moved. In its final version, ABE can travel in a straight line without even a wobble.

ABE. This robotic research vessel will be able to stay deep in the ocean for months, operating entirely on its own, collecting information with its sensors and cameras. Using a network of sound beacons, ABE will also be able to navigate to a precise set of sites on a regular schedule, letting scientists compare and observe changes that occur over a period of time. In between its rounds, ABE will shut down to save energy. At the end of its mission, a research vessel will return to an ABE site and signal it to rise to the surface so it can be picked up and the data studied.

While the technology hasn't yet been developed to make it possible, scientists hope ABE will one day be able to transmit information and video images directly to a satellite. This would let scientists study the seafloor as if they were actually there. The transmission could also be shared simultaneously by scientists in different countries working together.

The Scorpio *is a Remotely Operated Vehicle (ROV). The two arms you can see on the front are capable of lifting 250 pounds (112.5 kilograms) each.*

21

What's the Seafloor Like in the Depths?

It's Rugged

The seafloor has sharp jagged peaks and steep-sided canyons—a much more rugged terrain than what's usually found above sea level.

A system of mountain ranges that stretches about 40,000 miles (64,000 thousand kilometers) extends through all the oceans. Along the crest of some parts of this system, there is a central valley called the rift valley. This valley region is noted for its many active volcanoes. Sea mountain peaks that poke above sea level are islands. If a peak erodes away to a submerged flat-topped seamount, it's called a guyot.

The tallest mountain and the deepest trench on earth are in the ocean. Mauna Kea (33,476 feet / 10,042.8 meters) on the island of Hawaii is the world's highest mountain if measured from its base on the seafloor. The Marianas Trench (36,198 feet / 10,859.4 meters deep) in the western Pacific is the deepest trench. Compare that depth to the Grand Canyon, which is only 7,000 feet (2,100 meters) at its deepest point.

WHY IS IT RUGGED?

To find out why land forms at the bottom of the ocean are more rugged, try this activity. Working outdoors so it won't matter if you're messy, pile sand or soil up into a mountain with a pointed top. You may need to add a little water to make the sand or soil hold its shape. From a height of about a foot (0.3 meters) above the hill, pour on a glass of water. See how the water washes away some of the mountain and makes it rounder? Deep underwater, mountains aren't exposed to rain's weathering force. And they aren't attacked by wind or ice—two more forces that tend to round off mountains on land.

This diagram shows the general profile of the seafloor.

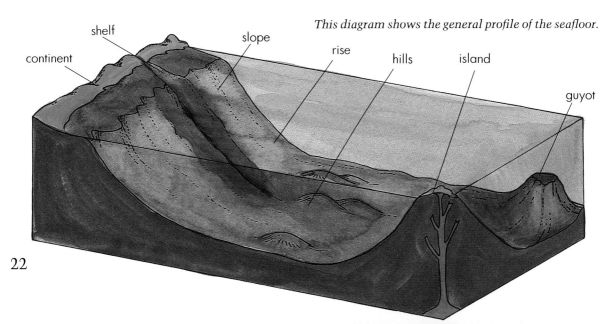

continent   shelf   slope   rise   hills   island   guyot

Even though all of the earth's oceans are connected, they are divided into four main areas. The Pacific Ocean contains the greatest volume of water and has the greatest average depth. Match each of these ocean features to the number on the map: (A) Hawaiian Islands; (B) Marianas Trench; (C) Mid-Atlantic Ridge. Turn the book upside down to check yourself.

It's Changing

It may surprise you to learn that the oceans are changing size. The earth is rather like an egg with a cracked shell. At the center of the earth there is very hot core. Surrounding that is the mantle, which is still hot enough for the rock material to be a thick, taffylike liquid. The crust, or outer layer of solid rock on which you live, is made up of about twenty separate pieces called plates that fit together like pieces of a puzzle. These plates are outlined on the map above so you can see them more clearly.

A-3, B-1, C-2

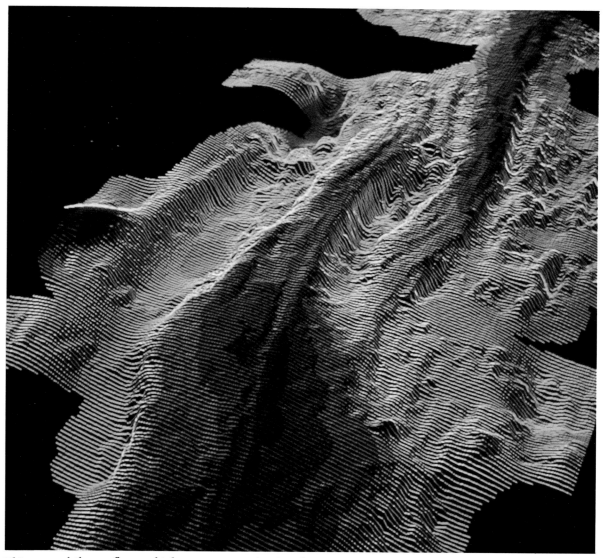

This map of the seafloor, which was generated using the Sea Beam system, shows a view of the midocean ridge in the east Pacific—an area of active spreading between plates. It's color-encoded to help geologists interpret the shape and structure of the seafloor.

Scientists believe that heat from the earth's natural radioactivity causes the molten material to circulate under the plates, making them move very slowly. As the plates move apart, the seafloor spreads and molten lava flows out, building up in some places to form new mountain ridges. Where plates run into each other and one plate slowly sinks under the other, deep trenches are formed. By determining the age of rocks along the plate boundaries, scientists have estimated that the seafloor of the Pacific is spreading faster than the seafloor of the Atlantic. But the rate at which this movement occurs is believed to vary from one area to another. At its fastest the seafloor in the Pacific is still only spreading about 2 inches (5 centimeters) per year.

There Are Hot Spots

In addition to volcanic activity along spreading plates, scientists have also discovered hot spots. These are places where superhot molten rock melts a hole in the crust. Then lava pours out, slowly building up a mountain that may become an island. That's what happened in the Pacific Ocean, creating the Hawaiian Islands. First, a volcano formed an island. Over time, the slowly moving plate changed position. The hot spot created another volcanic island, and then another, and another. The island of Hawaii is the most recent island in the chain, but it's not the last. Just 28 miles (45 kilometers) east of Hawaii's southernmost tip, a new volcano named Loihi is erupting. Although it already rises more than 16,000 feet (4,800 meters) from the seafloor, scientists estimate Loihi will need to grow an additional 3,000 feet (900 meters) before it appears above water. At the rate it's currently growing, travel agents won't be booking vacations on Loihi for at least another ten thousand years.

So they won't miss anything that happens as Loihi develops, scientists have stationed HUGO (the Hawaii Undersea Geo-Observatory) to operate around the clock. Video cameras on HUGO record eruptions while other instruments measure water temperature and test what chemicals are released into the water.

This computer-generated image of Loihi, the newest volcano developing in the Hawaiian Island chain, was created using Sea Beam data.

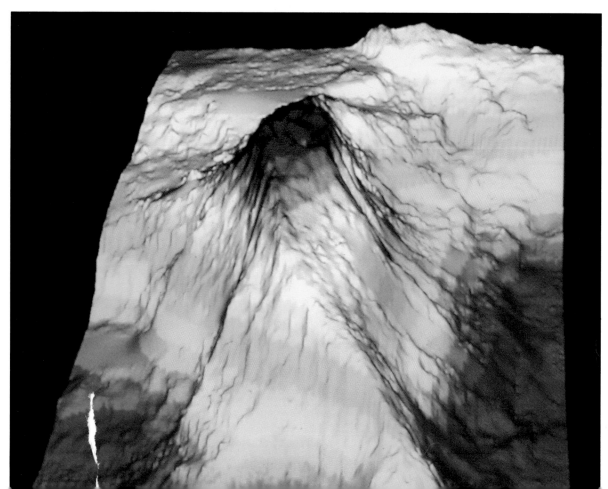

It's Mainly Covered with Sediment

Although the buildup of sediment is thickest close to the continents, even the deep seafloor far from land is covered with this loose accumulation of particles. Here's a look at the main types of materials that settle in the depths.

Land-based Sediment Weathered gravel, even tinier fragments of rock and sand, and volcanic ash are carried into the sea by rivers, wave erosion, and glaciers. The thickest layers of land-based sediment are found where big rivers, such as the Ganges in Asia, enter the ocean. Bigger, heavier particles are deposited closer to their source than tiny particles.

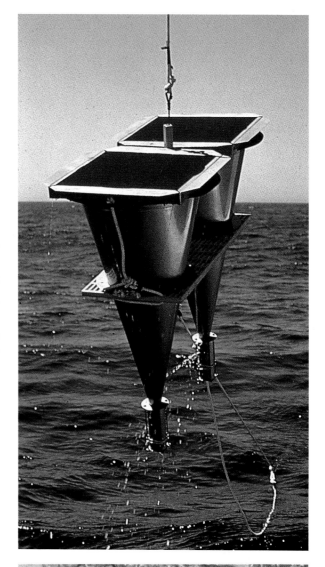

Top: This is a sediment trap. Scientists use these to catch sediment as it sinks to the seafloor. Because the particles settle to great depths very slowly, these traps may be left for months or even years before being picked up. By studying the material inside the trap, scientists learn about the types of particles settling in a particular area and can determine the rate at which the particles are drifting down.

Bottom: Tiny grains of clay minerals are freed when rocks weather and break down on land. Too light to settle quickly to the bottom, the grains of clay are often carried by currents to the deepest parts of the ocean before they begin to drift down. Strong desert winds also carry fine-grained sediment far out into the deep sea before the particles drop into the water and slowly sink to the seafloor.

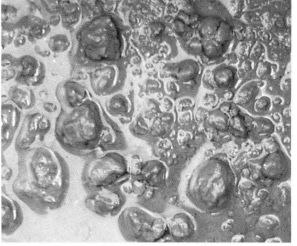

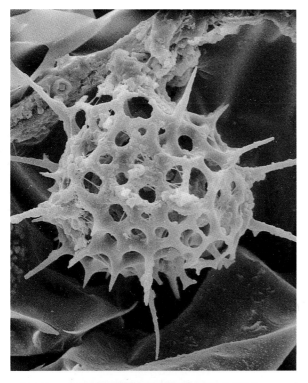

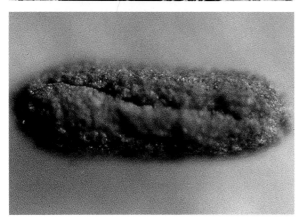

Oozes Sediments largely made up of the remains of plants, shells, teeth, and bones are called ooze. About half of the deep-seafloor is covered by an ooze that is made up mainly of shells from tiny marine snails, one-celled animals called foraminifera, and the hard remains of one-celled plants called coccolithophores. This type of ooze is rarely found in the deepest parts of the ocean, though. At great depths, there is a high concentration of carbon dioxide, the gas animals breathe out, and this gas combines with the water to form a weak acid that dissolves the tiny seashells.

WATCH IT SETTLE

To discover why the different-size sediment become sorted out, try this. Fill a quart jar full of water and dump in a handful each of the following: coarse gravel, sand, and powdery clay. Stir well. Then watch as the sediment settles. You'll discover that the heavier gravel settles first, then the sand reaches bottom, and finally the powdery clay floats down.

Top: This is the hard, glasslike remains of a tiny plant that lives at the ocean's surface. It's commonly found in ooze covering the seafloor in the Antarctic.

Center: This is the hard remains of another one-celled plant that lives at the surface of the ocean. After the plant dies, the buttonlike segments come apart and settle to the seafloor.

Below: This is a fecal pellet, the waste of an animal living in the sea. Pellets like this drift down, forming an important food source for some of the animals living in the ocean depths.

27

STACK A SNACK

You can see for yourself how such layers form. First, mix one cup of boiling water with a 3-ounce (61-gram) package of lime Jell-O. Stir until completely dissolved. Spoon about half a cup into a metal measuring container and set this in the freezer compartment for a few minutes. Then pour the partially chilled Jell-O into a two-cup drinking glass and set it in the freezer. Next, mix a cup of boiling water with a 3-ounce (61-gram) package of cherry Jell-O. As soon as the lime Jell-O layer has set, spoon half a cup of the cherry Jell-O into the metal measuring cup and place it in the freezer for a couple of minutes. Then pour the cherry Jell-O into the glass on top of the lime Jell-O. Return the glass to the freezer for five minutes or until the cherry Jell-O has set. Because the remaining Jell-O is no longer boiling hot, half a cup of lime Jell-O can be poured directly on top of the cherry. Repeat, adding layers, until the glass is full. Look at the glass from the side to view your sediment sample. Finally, for a taste treat, leave the Jell-O in the freezer for two hours and eat.

This section of a core sample lets you take a close look at the sediment layers on the seafloor. Oceanographers describe investigating a core sample as reading a book about the earth's past. The different-looking bands are caused by a change in the type of sediment or a change in the rate at which the sediment was accumulating.

Scientists estimate that in the deep oceans sediments build up at a rate of only 0.2 to 0.4 inch (0.5 to 1 centimeter) every thousand years. Along the coast, the sediment accumulates at a much faster rate, especially where big rivers reach the ocean. The Ganges, the Yangtze, the Yellow, and the Brahmaputra rivers—all in Asia—contribute the most—more than a fourth of all the world's land-based sediment each year. Based on core sample studies, geologists estimate that the seafloor sediments represent about 175 million years of the earth's history.

The layers of sediment that have stacked up on the seafloor provide valuable information about what the earth was like in the past: climatic conditions, periods of volcanic eruptions, ice ages, and variations in sea level. This information helps scientists model how the ocean responds to changes. Then they can predict what might happen in the future if, for example, there is continued global warming.

Core libraries are important. They hold samples scientists will be able to use in the future to answer questions that haven't even been thought of yet. These samples also let future scientists reconsider past theories as technology provides new tools for use in analyzing the sediment.

This is the core library at Texas A&M University. It stores about 75,000 samples from the Pacific and Indian oceans. A smaller core library at Scripps Institution in California also stores samples from the Pacific and Indian oceans. A library at the Lamont-Doherty Geological Observatory in New York State stores samples from the Atlantic Ocean.

What Are Deep-sea Animals Like?

Some Look Weird

Living where it's very dark and cold and the water pressure is extreme, the animals that make the ocean depths their home have developed some unusual features. Even those that look like monsters are small, though rarely more than a foot (0.3 meter) long.

Growing big takes a lot of energy, and food is scarce in the ocean depths. These creatures also breathe slowly and move slowly unless making a lunge to catch something to eat.

Bet you can guess why scientists named this fish the deep-sea swallower. Its enormous mouth is hinged to let it open as wide as possible—wide enough to swallow prey three times larger than itself. Its funnel-shaped throat leads into a tapering body that ends in a whiplike tail. ➤

◄ *The viperfish has special glowing organs inside its mouth to attract attention. When it gets close enough to the hatchetfish it's pursuing, the viperfish will lunge, snagging dinner with its long, sharp teeth.*

Many Deep-sea Animals Glow in the Dark

In fact, nearly three-fourths of all the animals living in the deepest parts of the ocean glow. Some animals glow to attract a mate. Others use their light to keep track of animals swimming with them in a group. Still other deep-sea animals glow to lure prey close enough to catch.

There are two main ways that these deep-sea glowers produce light. Some produce special chemicals that emit light in the presence of another triggering chemical or in the presence of oxygen. Others have saclike spots containing colonies of bacteria, tiny one-celled living things, that produce this same sort of chemical reaction to generate light. Either way, this light, just like the light of fireflies, is cool and doesn't burn. Since bacteria produce light all the time, animals that depend on the bacteria usually have some means, such as a muscular shutter, to block the light when they don't want to be seen. Animals that glow on their own usually only light up when they're disturbed.

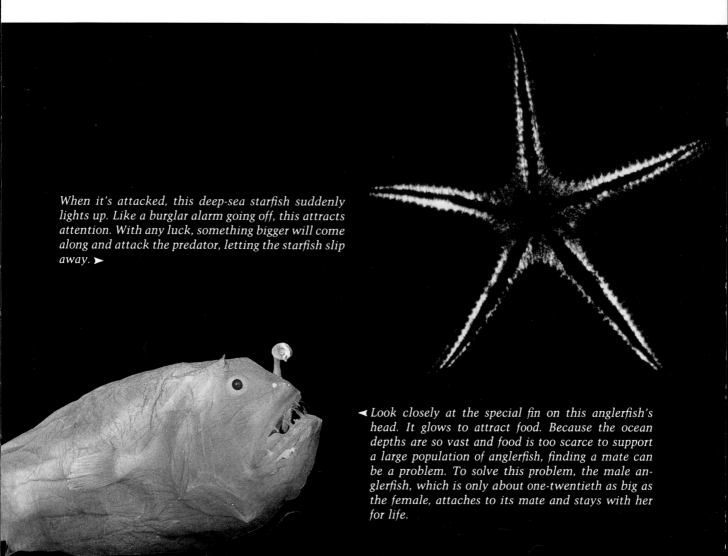

When it's attacked, this deep-sea starfish suddenly lights up. Like a burglar alarm going off, this attracts attention. With any luck, something bigger will come along and attack the predator, letting the starfish slip away. ➤

◄ *Look closely at the special fin on this anglerfish's head. It glows to attract food. Because the ocean depths are so vast and food is too scarce to support a large population of anglerfish, finding a mate can be a problem. To solve this problem, the male anglerfish, which is only about one-twentieth as big as the female, attaches to its mate and stays with her for life.*

How Are Deep-sea Animals Collected?

Sometimes scientists aboard a submersible collect deep-sea animals to bring back to the surface so they can examine them up close. The tool they use to do this is called a slurper.

Even animals that are successfully collected can't tolerate being taken from their natural habitat. For example, scientists collected one deep-sea animal that looked like a yellow dandelion. It was really a colony of small animals surrounding a central buoyant gasbag. When brought to the surface where pressure was much less than on the seafloor, this animal quickly fell apart. Some white crabs that live at great depths have been successfully brought to the surface for study by keeping them in a special container where the pressure is just the same as what they were used to—250 times what it normally is on the earth's surface.

This picture, looking out the front of a submersible, shows the slurper attached to the manipulator arm. The slurper works like a vacuum cleaner, with whatever is sucked in going into a bucket. The bucket is on a special rack under a lid that keeps what's inside from floating out. The rack turns the way a lazy Susan does, so each new slurped-up sample can go into a separate bucket.

COLLECT DEEP-SEA CREATURES

To see how a slurper works, fill the sink about half full of water. Drop in a few raisins, some unpopped kernels of popcorn, and a soda cracker. Hold the cracker underwater for a moment and then place a penny on it to hold it on the bottom. Use a turkey baster to suck up the specimens. To apply suction, you'll need to squeeze the bulb before you insert the tip into the water and keep squeezing until you touch the item you want to suck up. Then ease up on the bulb. Try to pick up a raisin and deposit it in a cup or glass. Then try to pick up just one popcorn kernel. Try to collect as much of the cracker as you can in one try.

Based on this investigation, what do you think are some of the problems of collecting animals with the slurper? Imagine what it would be like to try and catch an animal that was moving. Can you see why some animals could be seriously damaged by this collection process?

Minerals deposited around the deep, hot vents often build up irregularly shaped chimneys that may be as much as 8 feet (2.4 meters) tall.

Some Have Adapted to Living Around Deep-sea Hot Springs

While much of the deep-sea floor is as unfriendly to life as a cold dark desert, there are oases, special places where animal life thrives. Because so little of the deepest parts of the ocean have been explored, no one knew these unique communities existed until 1977. Imagine how exciting it was for the scientists who found the very first oasis.

They were on board the *Alvin*, 8,000 feet (2,400 meters) down near the Galápagos Islands in the Pacific Ocean. The first clue they had was when a temperature sensor signaled an increase in water temperature. Minutes later, they spotted what looked like black smoke and strange-looking animals, most of which had never been seen before.

Above: Except that they're white, the crabs that live near the vents look pretty much like crabs anywhere.

Left: These strange-looking animals are called tube worms. Some are as much as 6 feet (1.8 meters) tall. The red top of the worm is a spongy plume containing many tiny blood vessels. Here oxygen and a gas coming out of the vent called hydrogen sulfide are filtered from the seawater. Bacteria living inside the worm use these gases to make food for themselves and for the worms. Because this food doesn't need to be eaten or digested, the tube worms don't have mouths or a digestive system.

Alvin's crew had discovered a deep-sea hot spring called a hydrothermal vent. Such hot-water vents form when cold seawater seeps down into the crust through cracks in the seafloor where molten rock material is close to the surface. The heated water circulates back up through the crust, dissolving minerals, such as iron, copper, and zinc, and carrying them along. When the hot water reaches the seafloor once again, it shoots out like a fountain, and the dissolved minerals form crystals. Because the crystals make the shooting water look like smoke, scientists have named these hydrothermal vents smokers. Black smokers have mainly dark-colored minerals, such as iron sulfide. White smokers have mainly light-colored minerals, such as zinc sulfide.

Although the bigger forms of hydrothermal vent life, such as tube worms and white clams, are the easiest to spot, the most important are the tiny bacteria because they are the basic source of food. Near the surface of the sea and on land, green plants use the sun's light energy to produce food through a process called photosynthesis. Hydrothermal vent bacteria use the sulfides and other chemicals spewing from the vent combined with oxygen from the seawater to produce energy. They use this energy in a process called chemosynthesis to produce the food they need to grow and reproduce. The bacteria then become the basis of a food chain that supports many different kinds of animals. Some, like the tube worms and giant clams, have bacteria living within their bodies to supply food. Others graze on the mats of bacteria around the vents. Still others eat the animals that eat the bacteria.

Although marine biologists, scientists who study life in the sea, have learned a lot by observing the vent animals, they still have a lot of unanswered questions. For example, since these animals only live around hydrothermal vents and the hydrothermal vents are often long distances apart, how do the animals spread to new vent sites?

Right: What looks like a blurry streak is a jet of superhot water. Among the animals living around it are palm worms, named because their shape is similar to a palm tree, gray-and-pink scale worms, and giant tube worms with their red, plumelike tops.

Below: This vent animal is called a Pompei worm. It lives near a smoker chimney in a tube that it forms from minerals in the water. It's believed that the worm uses its tentacles to consume bacteria, but no one yet knows much about this animal's life.

These are the backbones—each about a foot (0.3 meter) long—of the whale skeleton that was first spotted in 1987. Scientists believe this was a blue or fin whale about 68 feet (20.4 meters) long. In the close-up of one of the ribs, you can see mussels and limpets—two of a number of different kinds of animals that make their home on the whale bones.

Some Have Adapted to Living Around Dead Whales

It may surprise you to learn that the body of a dead whale is also a sort of oasis in the ocean depths. Actually, no one was sure exactly what happened to whales after they died until researchers aboard the submersible *Alvin* discovered a whale skeleton on the seafloor off the coast of southern California. On closer inspection, they discovered a whole community of animals, similar to the ones found around hydrothermal vents, living around the bones. They also discovered that here, just as at the vent communities, the basis of the food chain was a type of bacteria capable of using chemical energy rather than sunlight to produce food. Since the vents were usually the source of the needed chemicals, scientists wondered what supplied the bacteria at the whale skeleton.

Further investigations revealed yet another kind of bacteria. These bacteria decomposed the whale's flesh and later the oil in the bones, and in the process they gave off the chemicals needed by the other bacteria to produce food.

Of course, as is so often true, what scientists learned about the animals living on the whale skeleton only made them think of new questions. How quickly do deep-sea colonies start after a dead whale sinks to the seafloor? How long is a dead whale's body able to support life? Scientists have returned several times to the whale bones, observing, taking photos, and collecting specimens to study in laboratories on shore. They hope to find answers to these questions and to think up new ones.

How Are Ocean Depth Discoveries Useful?

Shipwrecks Offer Peeks at the Past

This is the bottom of the USS *Monitor*, the Civil War ironclad ship that sank in a storm. Archaeologists, scientists who study ancient remains to learn about the past, are now able to search even deeper for shipwrecks and they don't even have to go to the seafloor themselves.

Remotely Operated Vehicles (ROVs) carry sonar and cameras into the depths. Once a wreck site is discovered, the ROV moves over and around the wreck, guided by a pilot on a surface ship. Cameras and three-dimensional scanning sonar give archaeologists an overview. Then special electronic still cameras able to detect details in the faintest light provide still more information. Finally, computers process all the data to improve the images. The detail in such images is often so good archaeologists can identify and date artifacts. Even better, it's now possible for archaeologists to view live video images transmitted from a ROV to the surface over fiber-optic cables. This makes it possible for them to take a closer look at one part or to redirect the ROV to survey an entirely new area based on what they're seeing at that moment.

The wide band outlining the side of the sunken ship is what's left of the thick metal armor that covered the Monitor's *sides and slowly rusted away after it sank. The ship is upside down, and the bump at the upper left is all that remains of the ship's gun turret.*

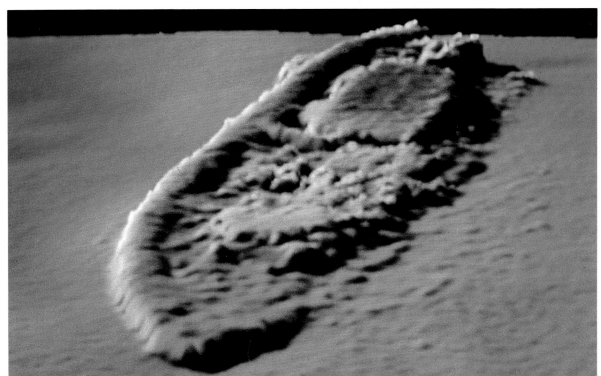

Oil and Gas Supply Energy

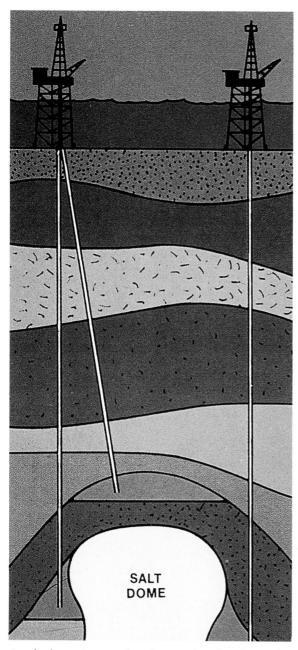

A salt dome forms when layers of rock salt are deposited and later buried by heavy overlying layers of sediment. Then pressures within the earth cause the rock salt to rise in columns, deforming the overlying beds. Migrating oil and gas accumulate between the particles, such as those of sandstone, in layers adjacent to the rock salt.

When the remains of countless plants and animals that have piled up over long periods of time are compressed by the weight of layers of sediment above them, these remains change chemically to form oil and gas. Oil and gas don't exist in big pools. Instead they fill up tiny spaces between grains of sand or rock particles, and they flow slowly toward the surface unless trapped by rock, such as shale, where the particles are tightly packed together. Then the oil and gas are likely to spread horizontally unless trapped by another impenetrable mass. Caught between two impenetrable layers, oil and gas deposits remain trapped until something happens to release them.

Unless some oil has reached the surface and is seeping out, scientists have to figure out where conditions are right for oil and gas to be trapped. Drilling into the seafloor is expensive and difficult, so companies only want to take the risk if they are likely to find oil. Special tools help. A seismograph was developed to record earthquakes, but it can also be used to record information about the structure of the seafloor. A big air gun creates a blast of energy that moves downward into the sediment and then bounces back to be recorded on the seismograph, revealing different layers below. Core samples show if there is porous rock capable of holding a reservoir of trapped oil. And a special device called a sniffer is pulled across the seafloor to check for traces of oil that may have seeped into the water. Computers process all the clues and construct models to help scientists analyze the proposed well site. Finally, though, an exploratory well has to be drilled to find out if there really are oil and gas trapped in the seafloor.

Along the coast where the ocean isn't as deep, drilling is usually done from a platform with long legs anchoring it to the seafloor. In deeper water, drilling is done from floating platforms or drill ships. These drill ships are held on-site by constantly running both propellers and special thrusters—rather like treading water. One exploratory well was more than 7,500 feet (2,250 meters) underwater.

If the exploratory drilling finds oil, production platforms are constructed. Sometimes the wellhead is housed in a special dry chamber on the seafloor and workers go down to maintain and service the equipment in a capsule that serves as a deep-diving elevator. More often, though, ROVs perform the deep-sea tasks under the guidance of a worker sitting at a control panel on the platform.

Athough the oceans currently supply 25 percent of the world's oil and gas, production has been limited to the continental shelf and slope. But scientists think there may be oil and gas deposits even deeper, waiting for the technology to be developed that will make it possible to drill at great depths. Researchers also continue to work toward improving methods of oil and gas production to reduce chances of polluting the ocean environment.

The derrick in the center of this drilling platform ship is used to lower the bit and drill pipe. As the well deepens, a muddy mixture of clay, water, and other chemicals is pumped down the drill pipe and through openings in the bit. This helps cool and lubricate the bit as it drills through the sediment. And it prevents a sudden gushing blowout that could happen when the bit finally breaks through the caprock and frees the trapped oil and gas.

Other Minerals Are Future Resources

Besides oil and gas, there are other useful and valuable minerals on the seafloor.

• Sand-grain-size particles of pure calcium carbonate, called ooids, are found in warm, shallow waters in places like the Bahamas, where they can be scraped off the seafloor with large dredges. Calcium carbonate is used to make cement and as agricultural lime by farmers to improve their soil.

• Phosphorite occurs as lumplike nodules and flat slabs along the edge of the continental shelf as it slopes into deeper water. It's used in the production of soaps, explosives, fertilizers, and steel.

• Sulfur, iron, copper, and other metals have been found at great depths, deposited at the hydrothermal vents called black smokers. These are often found along ridges where crustal plates are separating.

As long as an adequate supply of minerals can be found on land, the cost and trouble of collecting the undersea minerals, other than calcium carbonate, have kept companies from trying to mine them. But there is one other deep-sea mineral source—manganese nodules—that intrigues scientists and makes people wonder how they might be collected.

This is a piece of rock from a black smoker. The gold-colored inner layer is chalcopyrite, also called fool's gold. Although no one is interested in collecting this, the vent fluid also deposits valuable minerals. Why do you suppose no one is trying to mine these minerals at this time?

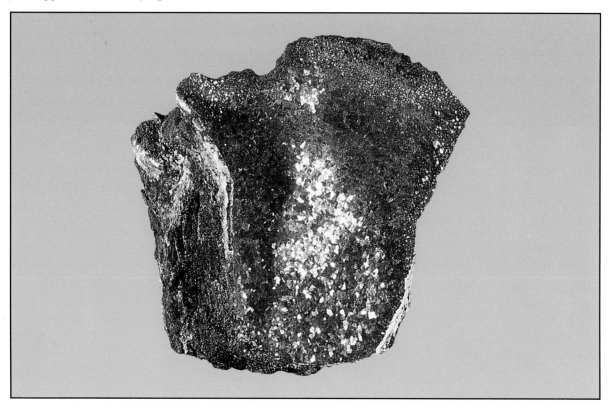

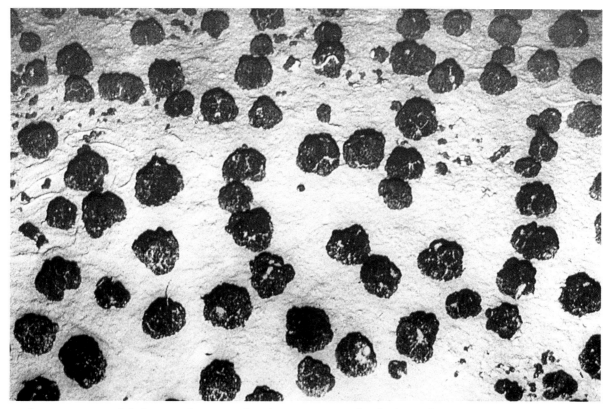

Each manganese nodule begins when particles of manganese oxide, the form of manganese present in seawater, collect on something hard, such as a grain of sand or a shark's tooth. Other metals dissolved in the surrounding water also collect on the manganese, slowly building up layer upon layer. No one knows for sure, but, because the nodules are nearly round, scientists believe currents keep rolling the lumps over and over as they form.

While manganese and iron are the most plentiful metals found in these nodules, these materials are plentiful on land too and are of little commercial value. Nickel is the most valuable mineral present in the nodules, and there are also significant amounts of cobalt and copper.

About twenty years ago, Canada, Germany, Japan, and the United States set up a joint venture to try to mine this deep-sea mineral resource. A test device somewhat similar to a vacuum cleaner with an oversized end and a 17,000-foot (5,100-meter) pipe was used to bring up the nodules to the research ship. As one scientist described the process, it was like standing on top of New York City's Empire State Building and trying to suck up peas on the sidewalk below with a very long straw. In other words, the effort was too difficult and too costly to make mining manganese nodules worthwhile.

Until land deposits of these minerals become scarce, researchers probably won't be interested in trying to develop the technology needed to do the job. In the meantime, though, can you think of a mining method that might be successful, using currently available technolgy? Or do you have an idea for a device using technology that hasn't yet been developed? Brainstorm. Draw a diagram to illustrate your plan. Be sure to consider the problems any deep-sea mining venture must face: great water pressure, lack of light, and extremely cold temperatures.

Animals Supply New Medicines

So far scientists don't know what medical resources animals from the deepest depths may offer. They already know, though, that some, such as *Trididemnum solidum* found at depths of about 120 feet (36 meters), provide new ways to fight diseases. Didemnin B, which is extracted from *Trididemnum*, has been found to be effective against a number of viruses, including the herpes virus that causes cold sores and the virus that causes yellow fever. It has also shown potential as an antitumor agent and testing continues to determine if didemnin B can be used to treat cancer patients.

With the help of submersibles, scientists can search much deeper for animals that might supply new medicines. Once a specimen is collected, a portion of it is immediately studied in a shipboard lab to see if any substances extracted from it will attack several basic strains of bacteria, yeast, or a type of fungus. If any of the tests are successful, the collection site is recorded and more samples are gathered for further studies at better equipped facilities back on shore.

Left: This is a type of tunicate called Ecteinascidia turbinata. *It's really a whole colony of animals living together; each of the small bumps is an animal. A substance obtained from this tunicate has shown signs of being effective in fighting leukemia and other types of cancer. Testing continues.*

Below: This is another tunicate called Trididemnum solidum.

Upwellings Mean Food and Jobs

Oceanographers who study currents and wave action are still trying to understand what causes mixing of water within the oceans. They do know that in some areas strong winds create currents that cause deep, nutrient-rich waters to rise to the surface. Conditions have to be just right for this to happen. So long as water density increases with depth, the water column is stable and very little mixing will occur. If the surface water becomes denser by becoming colder or saltier than the water beneath it, it sinks. Then the ocean's waters begin to circulate. Strong winds further encourage this mixing, and the deeper waters rise.

Such vertical mixing, called upwelling, is a vital form of recycling because the deep waters—especially along the coast—are

This sled was used in a project designed by researchers at Woods Hole Oceanographic Institution to find out to what degree salinity and temperature layering affect the mixing of midlevel ocean water. The study revealed that because of natural seawater layering, there was a good deal of horizontal mixing but very little vertical mixing.

rich in nitrogen and phosphorous. These chemicals are released when the remains of animals are broken down by bacteria. At the surface, nitrogen and phosphorous stimulate phytoplankton, tiny plants, to grow and produce food. The phytoplankton then become food for tiny animals, which become food for bigger sea animals, and so forth up the food chain. For countries like Peru, upwellings mean economic success. The phytoplankton in this case feed the fish that support Peru's fishing industry. Unfortunately, for reasons scientists don't yet understand, about every four to seven years, warm water flows south along the coast and blocks upwelled water from reaching the surface. Phytoplankton declines, and so does the fish population. The effect is devastating to Peru's economy.

This computer-generated map based on satellite data lets you see at a glance how well the phytoplankton are growing. Red- and orange-colored areas show the greatest mass of phytoplankton. Yellow and green areas show those with only moderate amounts. Blue and pink areas have very little phytoplankton. Black spots are areas for which no data was collected.

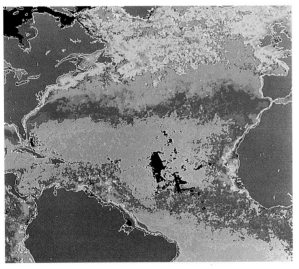

WATCH IT CIRCULATE

Here's an activity that will let you see for yourself how differences in temperature affect the mixing of water. You'll need a large, clear plastic storage box, three empty plastic film cans with snap-on lids (available at photo developing stores), ice water, very warm water, red food coloring, and twelve pennies.

Fill the plastic storage box nearly full of ice water. Place four pennies in each film can. Fill the cans almost full of hot tap water, add two drops of red food coloring, and snap on the lids. Set the film cans in a line down the middle of the storage box. Quickly remove the lids.

What happens to the hot water inside the film cans? What happens when the hot and cold water meet? Write a description of what you did and what you observed. Draw a picture to illustrate the circulation pattern that developed.

Repeat this activity, but this time fill the box with very warm water and fill the film cans with ice water. Write a description of this investigation too. Draw a picture to illustrate the new circulation pattern that developed and then compare it to your previous results. How is the circulation pattern different? In what ways is it similar?

Deep-sea Water Helps Generate Electricity

While Hawaii's warm surface water attracts tourists, it's the frigid water from the deep that's attracting businesses. Although it hasn't yet advanced beyond the research stage, scientists at the Natural Energy Laboratory of Hawaii Authority (NELHA), located at Keahole Point on the Kona Coast of the island of Hawaii, are pumping up deep-sea water to generate electricity. In tests, warm surface water is injected into a near vacuum, where it evaporates into steam. The steam is used to turn a turbine and produce electricity. Then cold deep-sea water is used to condense the vapor back into water to keep the system going. Since there are only a few land areas that have deep-sea water available nearby to produce electricity, scientists are also considering an extension to this system. The electricity created here would be used to obtain hydrogen from seawater. The hydrogen could then be shipped anywhere in the world and used as energy to generate electricity. One big problem so far, besides the initial cost of setting up the system, is finding materials that won't be ruined by being in constant contact with seawater.

These divers are checking one of the pipelines that reach down more than 2,000 feet (600 meters) to bring cold, very pure, nutrient-rich water to the surface.

Deep-sea Water Supports Marine Agriculture

These lettuce plants are being watered with the help of deep-sea water. Because Hawaii's air is naturally humid, frigid deep-sea water circulating through pipes causes the moisture in the air to condense on the pipes. The steady drip of this water is an effective watering system for the lettuce. This system is also good for supplying water for strawberries and flowers.

Cold deep-ocean water is also piped into tanks, making it possible to raise Maine lobsters and salmon that just naturally thrive in cold seawater. Like an upwelling, this water is also rich in dissolved nitrates and phosphates, so special aquatic farms are using this water to grow several varieties of seaweed. These seaweeds are nutritiously high in vitamins and protein and low in fat.

Into the Future

Now you know some of the ways people have already explored the ocean depths, what they've found, and how these discoveries are being put to use. Scientists estimate that no more than 5 percent of the sea bottom has been mapped in detail. That means 95 percent remains to be investigated. There is still a great deal to be learned about the processes that cause that vast volume of seawater to mix and circulate. And who knows what never-before-seen animals may yet be discovered. The technology making it possible to safely explore the ocean depths is being developed even as you read this book.

The bottom of the sea remains one of the last frontiers. You can help develop the equipment and tools that will make it possible to discover even more ways that the ocean depths can provide valuable resources. You may even want to become an ocean pioneer yourself.

Index